NEW AND IMPROVED C-4

Better-Than-Ever Recipes for Half the Money and Double the Fun

PALADIN PRESS
BOULDER, COLORADO

Also by Ragnar Benson:

Breath of the Dragon: Homebuilt Flamethrowers
Bull's-Eye: Crossbows
Eating Cheap
Gunrunning for Fun and Profit
Hardcore Poaching
Home-Built Claymore Mines: A Blueprint for Survival
Homemade C-4: A Recipe for Survival
Homemade Grenade Launchers: Constructing the Ultimate Hobby Weapon
Live Off the Land in the City and Country
Mantrapping
Modern Weapons Caching: A Down-to-Earth Approach to Beating the Government Gun Grab
The Most Dangerous Game: Advanced Mantrapping Techniques
Ragnar's Action Encyclopedia of Practical Knowledge and Proven Techniques
Ragnar's Big Book of Homemade Weapons: Building and Keeping Your Arsenal Secure
Ragnar's Guide to Home and Recreational Use of High Explosives
Ragnar's Homemade Detonators: How to Make 'Em, How to Salvage 'Em, How to Detonate 'Em
Ragnar's Ten Best Traps . . . And a Few Others That Are Damn Good, Too
Survival Poaching
The Survival Retreat: A Total Plan for Retreat Defense
Survivalist's Medicine Chest
Switchblade: The Ace of Blades

New and Improved C-4:
Better-Than-Ever Recipes for Half the Money and Double the Fun
by Ragnar Benson

ISBN 0-87364-839-0
Printed in the United States of America

Published by Paladin Press, a division of
Paladin Enterprises, Inc., P.O. Box 1307,
Boulder, Colorado 80306, USA.
(303) 443-7250

Direct inquiries and/or orders to the above address.

Contents

Warning

The procedures in this book and the resulting end products are extremely dangerous. Whenever dealing with high explosives, special precautions should be followed in accordance with industry standards for experimentation and production of high explosives. Failure to strictly follow such industry standards may result in harm to life or limb. Therefore, the information in this manual is *for academic study only*. Neither the author, publisher, nor distributors of this book assume any responsibility for the use or misuse of information contained herein.

Preface

These are indeed the times that try one's soul. On almost every hand we are beset by people who believe that, in spite of total, 100 percent international evidence to the contrary, government can find solutions to problems.

"You got a problem, let's pass laws and hire bureaucrats who will implement a program to solve it," the majority seem to be screaming.

Rather than recognizing basic human nature for what it really is, independent, self-sufficient, freedom-loving Paladin readers are often lumped together by political bosses as part of the problem. Prestigious publications, such as the *Wall Street Journal* and *The Economist*, have attacked Paladin for publishing information that until very recently was commonly known by every farm boy and was part of our ability to live peaceful, free, productive lives.

Free, unencumbered use of high explosives contributes a great deal to the strength of the American economy. Explosives allow an individual working alone to accomplish work that previously would have taken thousands of laborers and likely cost millions of dollars. Without explosives we could not quarry rock needed for roads, clear harbors, straighten rights of way, or even dig ditches efficiently. By using explosives farmers working all by themselves could clear and drain land on which to raise an abundance of food required by an expanding population. Explosives also assisted with controlling floods, securing lone rural homes, digging fence postholes, and keeping fascist politicians at bay.

It seems like only yesterday—but really was 24 years ago—that I ran into a huge, seemingly immeasurable basalt rock while building a road in northern Kenya. The rock appeared in the middle of our hand-dug right-of-way running along the side of a steep canyon. Completion of this particular three-mile stretch of road would cut 75 miles from the trip into town.

Construction of this hand-dug road in Kenya was temporarily halted when we ran into a basalt rock column shown in the top left.

Preface

Our crew of about 50 day laborers cut down around the rock; we were hoping we could loosen and roll it off the steep side of the canyon. But the more we dug, the bigger and more monolithic the rock became. Rerouting the road upward would have cost us three weeks' work at a minimum.

My simplistic solution to the problem was to use about 10 pounds of 60-percent dynamite to mudcap the rock, shattering it down sufficiently that large pieces could be pried loose and rolled down into the canyon. Smaller chunks would be used to make a nice fill.

Our host in Kenya reckoned that it would be cheaper to buy a dozer and ripper to handle the rock than to find what was then $5 worth of explosives. He was very well hooked into the government, so we decided I should at least try talking to the local district commissioner (DC) to see if I could get permission to purchase explosives.

The DC was extremely sympathetic to our arguments and to the fact that completion of this road would provide great economic benefit. He also believed that I knew what I was doing with explosives and that the whole affair was a simple undertaking. However, he pointed out that under Kenyan law he was obligated to apply to the military for explosives, which then had to be hauled in special trucks on a 17-hour trip from Nairobi. I was informed it would take six months, at best, to pass necessary paper and take delivery.

He laughed good naturedly when I asked if I could just drive to Nairobi and pick up the necessary powder and caps. Not even the offer of a healthy bribe shook him loose. I was of a mind to try to make some explosives from ammonium nitrate fertilizer, but at that time I had not experimented sufficiently to know how to make caps. At any rate, even ammonium nitrate fertilizer was illegal in Kenya!

The DC recognized that economic development in his already backward country was suffering as a result of the

Following a six-month delay, the Kenyan road was finally completed by using an expensive bulldozer.

many petty rules and regulations, of which getting military permission to use explosives was only one. But he was unwilling or unable to change anything. Association with people using inherently suspicious commodities such as explosives was potentially more dangerous to his career than the alternative of doing nothing.

That was 24 years ago. I thought then that it was little wonder that Kenya was such a backwater, decreasing in economic ability, while the United States continued to grow economically. Thankfully, I naively thought, we do not have to endure this same heavy hand of government regulation in the United States.

But that was back when explosives were commonly available in country hardware stores and cost about 10 cents per pound!

Today, in the United States the only way many citizens can reasonably expect to acquire explosives is to make them from commonly available materials. Some states still have reasonable laws, but in general it has become expensively difficult to purchase simple explosives with which to take out the common boulder.

This book attempts in every regard to specifically detail formulas and procedures that effectively use commonly available materials to produce high explosives needed to do common work.

Introduction

Tens of thousands of faithful readers have purchased copies of *Ragnar's Guide to Home and Recreational Use of High Explosives*, *Ragnar's Homemade Detonators*, and *Homemade C-4*, and many have taken the time to write to me regarding their experiences with explosives. In all cases, I appreciate these letters and the information and comments contained therein.

One of the most frequently asked questions is whether there isn't something effective out there that one can use in place of horribly expensive and often difficult-to-find nitromethane. "We appreciate," the writers often comment, "the simplicity and relative safety of basic C-4, but we are not able to acquire nitromethane."

Another frequent question involves the manufacture of effective explosives from anything other than nitromethane and ammonium nitrate. "Are there other formulas which are as effective?" they write.

"Many formulas, including yours, are included in government manuals," readers claim, "but usually exact details are sketchy as to how to successfully put these materials together. In some cases these formulas simply don't work." Mixing nitromethane with fine, hardwood sawdust is an excellent case in point. It's a waste of good nitromethane, and that's all. A number of skilled explosives people have tried but cannot make this mixture detonate. It is important, readers say, to accurately learn what will and will not explode.

Unconfirmed reports from suppliers of fuel for small model engines, which contain varying percentages of nitromethane, indicate that about three years ago one of the principal nitromethane manufacturing plants in California burned down. Flyers of model planes and drag racers suffered a real tightening of the nitromethane supply as a result. Months went by while California environmentalists fought against rebuilding the plant, but finally the manufacturer is reported to be back on line, and supplies are flowing again. Prices of nitromethane may come down, and its availability may become more general. On the other hand, government regulators could prohibit widespread sales of nitromethane, which some observers claim is already the case.

The challenge for this book, then, is to come up with simple, workable, readily available, generally prudent alternatives to basic C-4. These must use inexpensive, unregulated chemicals that average, knowledgeable citizens can find even if they live in large cities. These components must be relatively easily assembled in an emergency situation that requires high explosives. Government documents relative to recent interest in improvised explosives suggest that most officials consider homemade explosives to be too mild and erratic to be of much value. We must also come up with materials that refute such drivel.

In all cases, the assumption is made that no reader will actually assemble any of these formulas unless faced with an absolute emergency. All the procedures and formulas that follow range from dangerous to *very* dangerous. Because an explosive worked safely once is no guarantee that it will not bring tragedy to the novice the next time he experiments with it.

Recall Che Guevara, who during his early revolutionary days in Cuba, claimed to have lost "half the people I sent to producing explosives and explosive devices."

Those who have not at least read the three above-mentioned books or who have not had some formal training in explosives deployment should not attempt to deploy any of the explosives listed in this book, even in an emergency. It is always much safer to have learned to handle relatively stable, safe, easy-to-use military or commercial explosives before moving on to the erratic, inconsistent, home workshop ones set out in this volume.

In a related note, a great deal of detailed information regarding what does not work is included in this book. It is hoped that it will keep readers out of expensive and often dangerous dead ends.

During any testing process, it is desirable—even necessary—to keep all variables outside the test to a minimum. For this reason, most detonators used in these shots were commercial ones, rather than homemade.

Double-capped or booster charges, as will be repeated over and over, were often deployed. These are specially assembled primers consisting of a standard, fused cap and two one-inch segments of 50-gram detonator cord taped to the cap. One can easily make up double-strength homemade primer that, in an emergency context, could accomplish the same work, assuming an absolute emergency wherein everything must be improvised.

Boosted primers are often necessary to get some very sleepy but powerful explosives to detonate properly. Readers should not forget that all military explosives are much tougher to detonate than commercial ones.

Homemade C-4 is always the benchmark. Even though

Throughout this manual, the author makes the assumption that readers have all the information contained in the above Paladin *books or has formal training in explosives deployment.* Caution: *All of the procedures and devices contained herein are extremely dangerous!*

most other formulas are not as able as homemade C-4, please recall that the objective is alternative recipes composed of easily found, relatively cheap components. It's like Sears and Roebuck in the old days when they offered good, better, and best merchandise. Often, good did the job nicely at a price markedly reduced from best.

Explosives are necessary to operate a modern economy. It is hoped that readers won't have to use the following explosives to keep their economy afloat. But here they are—just in case they are the only game in town!

1 Explosives from Nothing

There are believable accounts regarding desperate people in occupied France or even New York City during World War II who constructed crude firearms in their home workshops. Guns produced under these circumstances had limited power, range, and accuracy but were entirely adequate for securing a real, operational weapon from the then enemy. In that regard, these often extremely primitive weapons had great value.

Unfortunately, no close analogy exists for homemade explosives. It would be nice if some relatively wimpy, weak explosive could be used to obtain a better grade of military explosive, but no one can tell me of an instance where this has actually happened. So far, any such tales are speculative, but you can always remain optimistic.

Many readers who have taken time to write me about explosives have asked for a dirt-simple, easy-to-make explosives formula that uses only the most common of compounds. Oddly enough, this is also a question frequently asked by newspaper and TV folks, who seem oddly fixated on the subject of improvised explosives.

Fine dust, evenly dispersed throughout a closed space can be extremely explosive, I always point out. Grain elevators are in constant danger of destruction from dust explosion. Many complicated and expensive measures are taken by people who store dried grain to ensure that such explosions are prevented.

On several occasions back on the farm, we had the opportunity to take down redundant buildings with explosives. The most effective technique was to vaporize gasoline by using a bug sprayer. A quart of gasoline atomized into a 15-x-35-foot building and shot with a cap really unhinged things. Hollywood demolitions experts create showy blasts that moviegoers like by enhancing their explosions with gasoline, but this is not what people want to read about.

A few very simple, relatively easy-to-formulate explosive mixtures exist, made from the commonest of materials. And

All homemade explosives are ranked on the basis of how they compare to commercial dynamite.

unlike dust and spray gasoline, these mixtures are controllable. Yet they are so wimpy that you could validly ask, "Why bother?"

The following recipe has explosive equivalents of about 30-percent dynamite. To the best of my knowledge, 30-percent dynamite has never been commercially produced, which may say something about this its utility. The material we are about to describe is not quite 40 percent but significantly more powerful than 20 percent, both of which have been assembled for the commercial market.

INGREDIENTS

Aluminum Powder

Extremely fine-ground, 300-mesh aluminum powder is the toughest and most expensive component to locate, but it is necessary for whomping up this first explosive. Automotive Alumaseal, as used to boost nitromethane/ammonium nitrate explosive, will work but not well, probably because the aluminum powder in Alumaseal is not fine enough. Explosives manufacturers require really fine aluminum dust.

Government-produced improvised explosives manuals, many of which contain B.S.-type formulations that no one

Commercial automotive Alumaseal can be used in homemade explosives, but it is not the best choice.

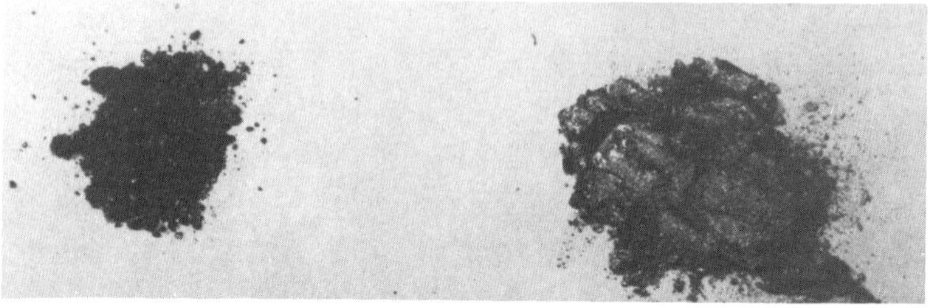

Alumaseal (on the left) will work to boost detonation velocities of explosives, but more finely ground 300-mesh aluminum dust (right) works much better. Finely ground aluminum powder is very dustlike; a little quickly permeates an entire explosive mixture.

seems to be able to make work, suggest purchasing powdered aluminum from full-service paint stores. Perhaps 20 years ago, when these books were first printed, you could purchase powdered aluminum from paint supply houses. Today, it is easier to jump through all the many hoops necessary to purchase aluminum powder from one of the many high school chemistry lab supply houses than to persuade a paint store to order powdered aluminum.

But some ever alert readers have discovered an excellent loophole in the availability blockade. In essence, yogurt and granola eaters now found in virtually every corner of the land have come to our rescue. These are folks who generally support local silk-screen shops, the little newly emerging businesses that produce short-run signs, T-shirts, and bumper stickers. People in all these shops have on hand, can order for you, or can tell you where to order finely ground, dustlike 300-mesh aluminum.

Aluminum powder is commonly supplied by national silk-screen supply houses. They also have bronze, gold, and blue powder, but only aluminum powder is pure metallic. Others contain copper-aluminum alloy or colored plastic. These

Finely ground 300-mesh aluminum powder is used in silk-screen shops where it can often be purchased for about $30 per pound.

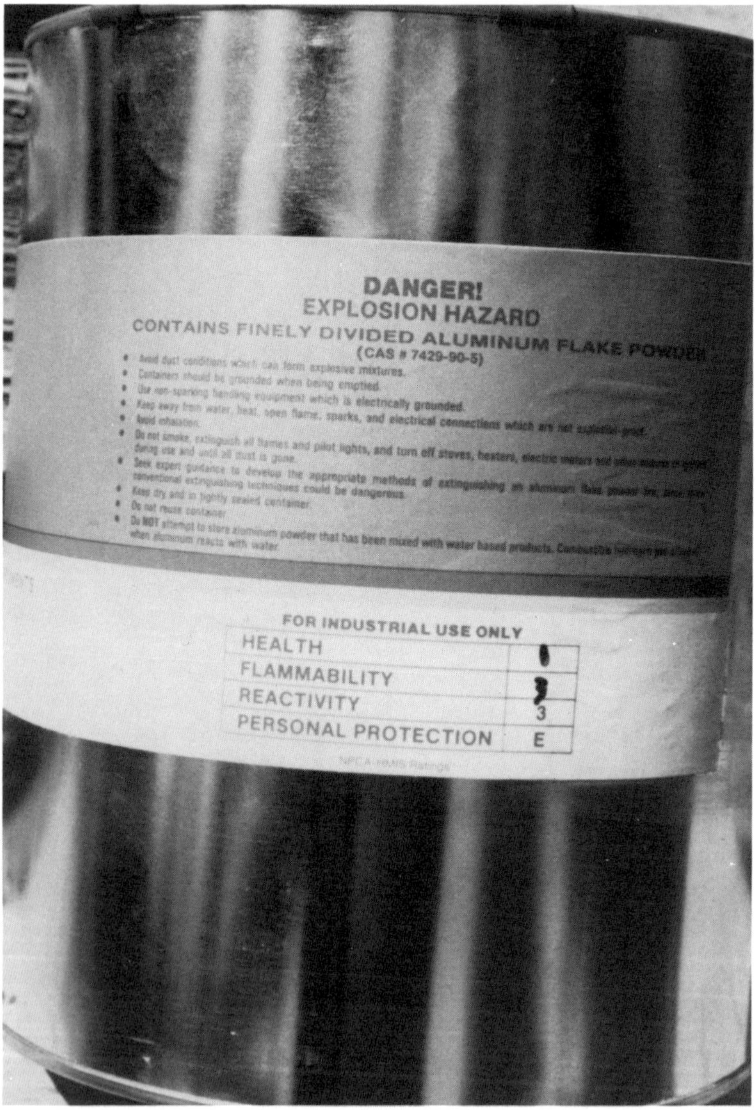

As indicated on the label, aluminum powder is considered to be highly reactive, flammable, and dangerous.

other materials, especially plastics, are completely unsuitable for the task ahead.

Armed with this knowledge, average citizens can purchase aluminum dust off the shelf in small to large cities virtually anyplace in the United States. Because the material costs about $30 per pound, and a pound lasts a long time in their line of work, most silk-screen shops carry relatively small amounts. In other words, these shops have the goods, but you must generally plan well ahead for a certain supply.

A national sign and screen print supply company has branches in 10 different cities, but listing its name and address here would guarantee that no one could purchase aluminum from any of its outlets. My point is that explosive makers who root around a bit at their local silk-screen shops will generally come up with acceptable materials.

Most silver-colored paint has a minimum of 15 percent of aluminum powder pigment base. Perhaps the petroleum vehicle would sensitize ammonium nitrate if it could be thoroughly mixed and molded before drying. However, this is an idea for a mixture that has not, to the best of my knowledge, ever been tested.

Nitromethane

The next necessary ingredient in terms of procurement difficulty is model airplane fuel. Makers must find the more expensive and obscure type containing 40-percent nitromethane. Not only must this fuel be the hugely more expensive 40 percent variety, but it must be reasonably fresh (less than one year old). Hobby shops often carry 15- to 25-percent nitromethane fuel but not usually the 40-percent variety.

You can either place a special fuel order with your local hobby shop or order by mail from Tower Hobbies, Box 9078, Champaign, IL 61826. Cost, not including shipping, will be about $30 per gallon unless you purchase from Tower, where near-wholesale prices prevail.

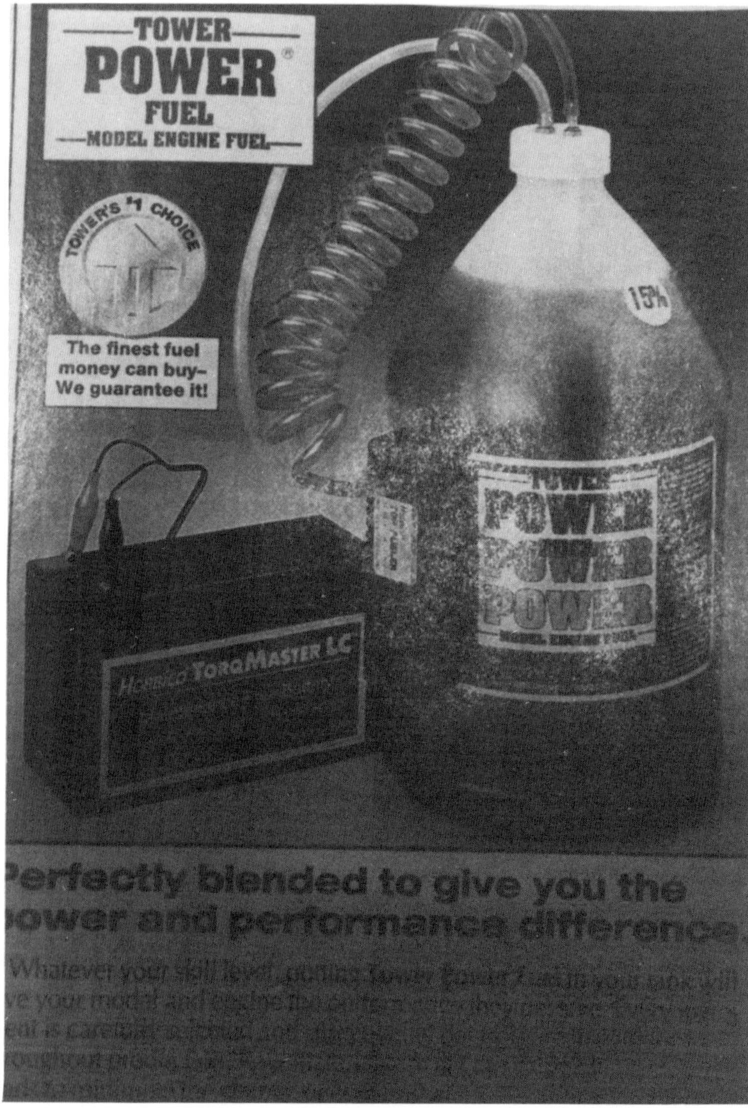

This model fuel contains 15-percent nitromethane, the highest concentration many hobby shops carry. Improvised explosive makers must use 40-percent nitromethane fuel, even if it must be special-ordered.

Use only fresh 34-percent ammonium nitrate when making any explosive. Even fresh material must be baked, cleaned, and dried with fresh alcohol.

Ammonium Nitrate Fertilizer

The next ingredient is fresh, dried, finely ground ammonium nitrate ($N_2H_4O_3$) fertilizer. As is true whenever you use ammonium nitrate, even the very smallest amount of moisture will spoil the explosive. Use only fresh ammonium nitrate purchased from any agricultural or garden supply outlet. Do not use old ammonium nitrate—it won't work.

Wash the clay coating from the ammonium nitrate with new alcohol. Finely grind the prills into extremely fine powder quickly, before humidity from the air destroys the powder. Readers must master this procedure (described in the homemade-C-4 book and video) before they embark on any clandestine explosives manufacture outlined in these chapters.

Ammonium nitrate attracts water from the air like a sponge. Therefore, the clay coating must be washed from the tiny prills before using. Cleaned, washed ammonium nitrate must be used immediately or stored in double-sealed plastic containers.

Common household flour is the principal ingredient in this improvised explosive.

Flour

The last ingredient for this very field-expedient explosive is common, household wheat flour of the type available anyplace other than Cuba and Ethiopia—and perhaps California after politicians read this book.

MIXING THE EXPLOSIVE

As in handling any explosives, procedure and caution count at least 99 points. Use the proper measuring equipment and proceed carefully as follows:

1. Measure 250 milliliters of flour into a glass or nonmetallic bowl. Other amounts can be made up using similar proportions, but by starting with 250 milliliters of flour, one can nicely fill a six-inch plastic cartridge.
2. Thoroughly mix in one tablespoon (30 milliliters) of finely ground aluminum. Use regular silk-screen-grade aluminum if available. If not, make do with coarser Alumaseal.
3. After the powdered aluminum is thoroughly mixed in, measure out three tablespoons (45 milliliters) of powdered, dried ammonium nitrate from a sealed plastic bottle and thoroughly blend it into the mixture.
4. Adding in the right amount of model fuel is somewhat subjective. In general, expect to use about 125 milliliters of 40-percent nitro fuel when starting with 250 milliliters of flour. The problem with recommending precise amounts of fuel arises as a result of this fuel's not really soaking into the flour.
5. Mix and knead the flour, ammonium nitrate, and powdered aluminum mixture. As the fuel is added, a rubbery, tough mixture will form. Add enough fuel so that you can almost squeeze a few drops of fuel out of the mass. Be very certain that everything is extremely well mixed. Most makers knead the mass like so much foul-smelling bread dough.

(As an aside, it has not been possible to reliably sensitize regular ammonium nitrate fertilizer using 40-percent nitromethane fuel to create an explosive. You would suppose that adding enough model fuel would do the job, but as a practical matter this has not been the case.)

An inexpensive set of plastic measuring spoons and a measuring cup marked in milliliters are essential for accurate measuring.

LOADING THE EXPLOSIVE

Loading this explosive, which does not seem to be shock or flame sensitive, is also critical. It must be loaded before the fuel dries. Unlike better, high-grade, high-velocity explosives, this material will not detonate with any enthusiasm unless it is tightly packed in strong cases.

One-and-one-quarter-inch-diameter, thin-walled plastic plumber's pipe cut into six-inch lengths fitted with standard end caps is ideal as a cartridge body. This plastic pipe will endure about 10,000 pounds of pressure, thus providing the containment necessary to achieve a blast.

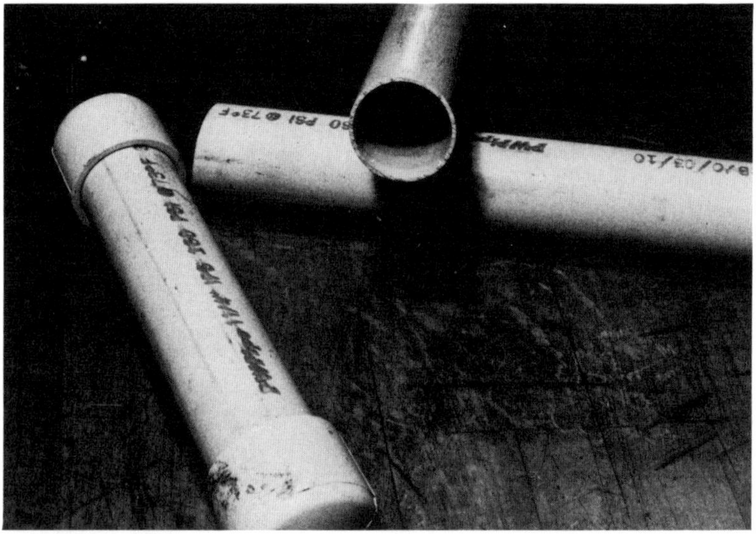

Other than the advanced C-4 outlined in Chapter 4, all other explosives discussed in this book must be tightly packed into strong containers such as these cartridges made from 1 1/2-inch plastic pipe.

Use a one-inch wooden dowel rod to firmly tamp the mixture into the plastic cartridge bodies. Do this a layer at a time till the mix is very solidly packed into the pipe. As far as can be determined at this time, this explosive is not particularly sensitive.

Place the second end cap on the full cartridge and allow the loaded mix to stand for at least four hours. This mix may become extremely sensitive after a longer period of time, but this is uncertain at this time. Nevertheless, beware of this and any other homemade explosive.

DETONATING THE EXPLOSIVE

Successful firing requires a boosted commercial cap or a double-strength homemade cap. Again, these bigger caps are sensitive and can do severe damage by reason of their increased size. Cap assemblies must be set down firmly in the middle of the cartridge to achieve successful detonation.

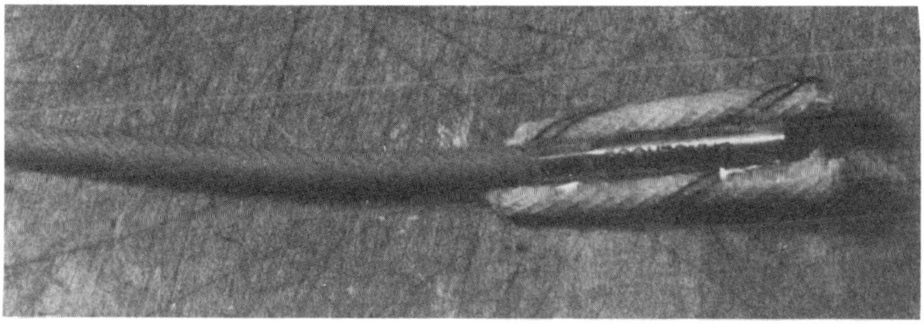

Most improvised and military explosives are sufficiently sleepy that they are best fired with boosted caps. They can be constructed by taping two segments of 50-gram Primacord to the cap or by using a double-strength homemade cap.

It is unknown whether these cartridges will propagate when fired one against another. You may be forced to load three-inch-diameter, 18-inch-length plastic pipe bodies to get a sufficient charge to do any real work using this material. However, using giant cartridges to do work with this explosive is probably not practical. But in keeping with our original criteria, we now have a workable recipe, formulated from dirt-common, easily acquired, cheap materials.

Readers should let me know if they find a true, practical application for this material

2 Great Potassium Chlorate Formula

Suggestions for making homemade explosives by adding nitromethane to fine hardwood sawdust or mixing 6-percent ammonia-based window cleaner with nitromethane share a common trait: they are all listed as being excellent improvised high explosives in government manuals. But it's like the old story—"I'm from the government, and I'm here to help you." It just doesn't work!

You can easily use up great quantities of expensive, often difficult-to-find nitromethane and blasting caps in figuring out that, in practical terms, these formulas are somebody's pipe dream. Results, as far as I can determine, are always zero.

This leads us back to more basic—and perhaps more easily found—materials with which to make explosives. In an aggregate sense, it is difficult to know whether readers can more easily locate supplies of nitromethane or potassium chlorate. Nitromethane is always available at drag strips, but potassium chlorate ($KClO_3$), it seems, is seldom available anyplace any longer. Perhaps our society has discovered other materials to use in place of potassium chlorate. Compared to the situation 20 years ago, it just ain't out there no more.

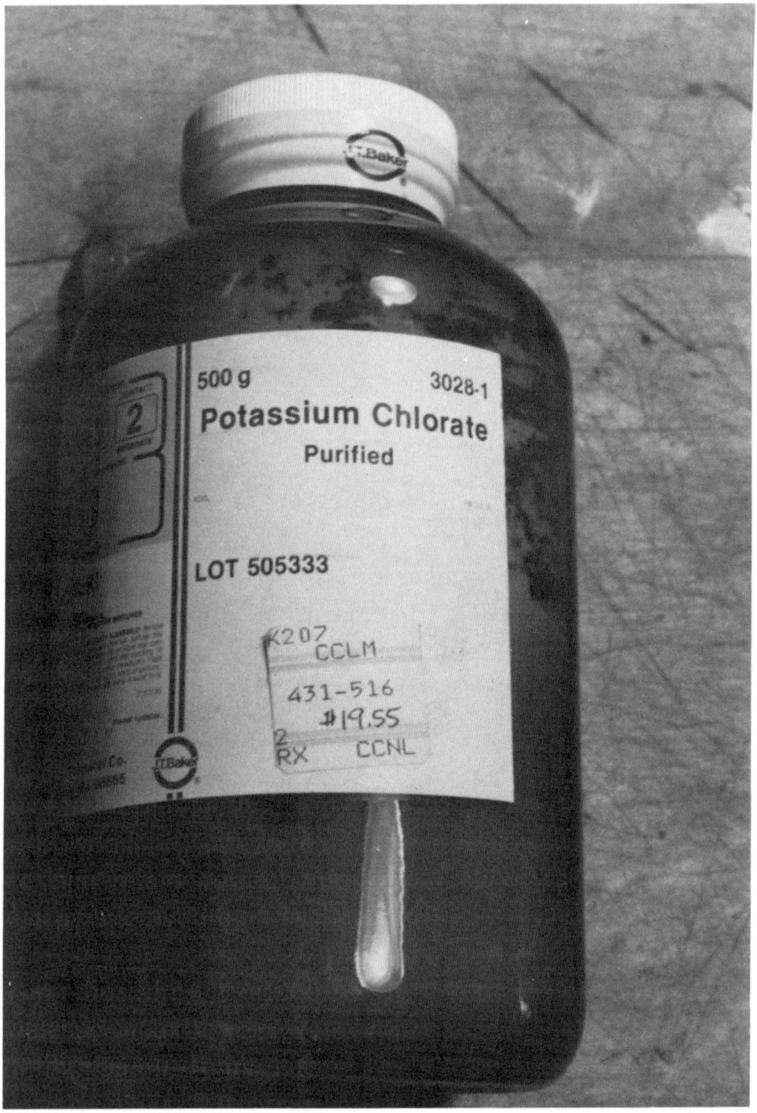

Once commonly available in virtually every drugstore, potassium chlorate is the basis for some fairly good homemade explosives. And when this material is finally found for sale, it is extremely expensive and offered only in small quantities.

Some small-town pharmacists, agricultural supply stores, and veterinarians can occasionally be persuaded to special-order and sell potassium chlorate, but this is far from certain—especially in quantities needed to make enough explosives to do some productive work.

The larger, chain-type pharmacies that currently are filling the land are no help for items outside their normal operating area. Either company policy prohibits the sale of chemicals, or people manning the counters do not know how—or want—to get such materials.

Most modern, chain-type drugstores do not carry or special-order chemicals such as potassium chlorate.

As a result, explosives using potassium chlorate as a base may not be practical for those with significant amounts of work to do. Anyone finding a reliable industrial source for this chemical should let others of us know as soon as possible.

However, there is an excellent, battle-proven recipe for an explosive based on potassium chlorate that is roughly equivalent to 60-percent dynamite. I must leave it to the readers' own initiative to discover how they can accumulate enough potassium chlorate to do any productive work.

Start with 200 milliliters of finely ground potassium chlorate. This material comes from the factory as very fine, sugarlike crystals or tiny prills. In this form it is completely, absolutely inadequate for making explosives. Making powdery potassium chlorate is quite difficult, but it must be done properly or results will be unsatisfactory.

Some makers use two flat boards to grind the potassium chlorate. Others employ a large mortar and pestle. Electric coffee grinders work poorly unless you keep individual batches very small. Potassium chlorate is simply too fine and dense in its natural state to grind properly in a small home-

In its original form, potassium chlorate is made up of granular, sugarlike crystals.

To create a reasonable good explosive, potassium chlorate must be finely ground into dustlike particles and mixed with petroleum jelly.

owner-type grinder. Dump the finely pulverized material into a large ceramic bowl for mixing.

Wash 20 milliliters of new, clean ammonium nitrate in alcohol to dry and to remove the kaolinite coating. Immediately, before it can again suck moisture out of the atmosphere, evaporate away the alcohol and mix the prills thoroughly in the potassium chlorate dust. If the last step is not practical at this time, seal the mixture in a tight glass or plastic jar, which is then resealed inside a plastic Ziploc bag.

As is frequently true, precise measurements are not possible; but thoroughly mix about 40 milliliters of warm petroleum jelly into the potassium chlorate-ammonium nitrate mixture. The petroleum jelly can be warmed to a more fluid state by submerging a closed jar of it in hot water.

Knead, chop, roll, and stir with a wooden stick until the

Regular over-the-counter petroleum jelly can be mixed with potassium chlorate to make 60-percent dynamite or with Bullseye *sensitized ammonium nitrate to make plastique.*

Potassium chlorate must be carefully and thoroughly ground into a very fine powder. There are no quick and easy ways to do this, but a mortar and pestle works as well as anything. You should grind only small amounts per batch.

petroleum jelly is fully incorporated into the entire mass of powder and prills. If 40 milliliters of petroleum jelly is not sufficient to pick up all of the powder, add a bit more, a teaspoon at a time. Thoroughly mixing this material is about as troublesome and time consuming as powdering the potassium chlorate.

In its preferred state, the mass should no longer be sticky (a sign that too much jelly was used) nor powdery (an indication that insufficient jelly was used) In general, the lump must be homogeneous and a little softer than a peeled banana.

Like the flour explosive in Chapter 1, this potassium chlorate material works best when tightly pressed into a rigid, tough cartridge container that will withstand considerable

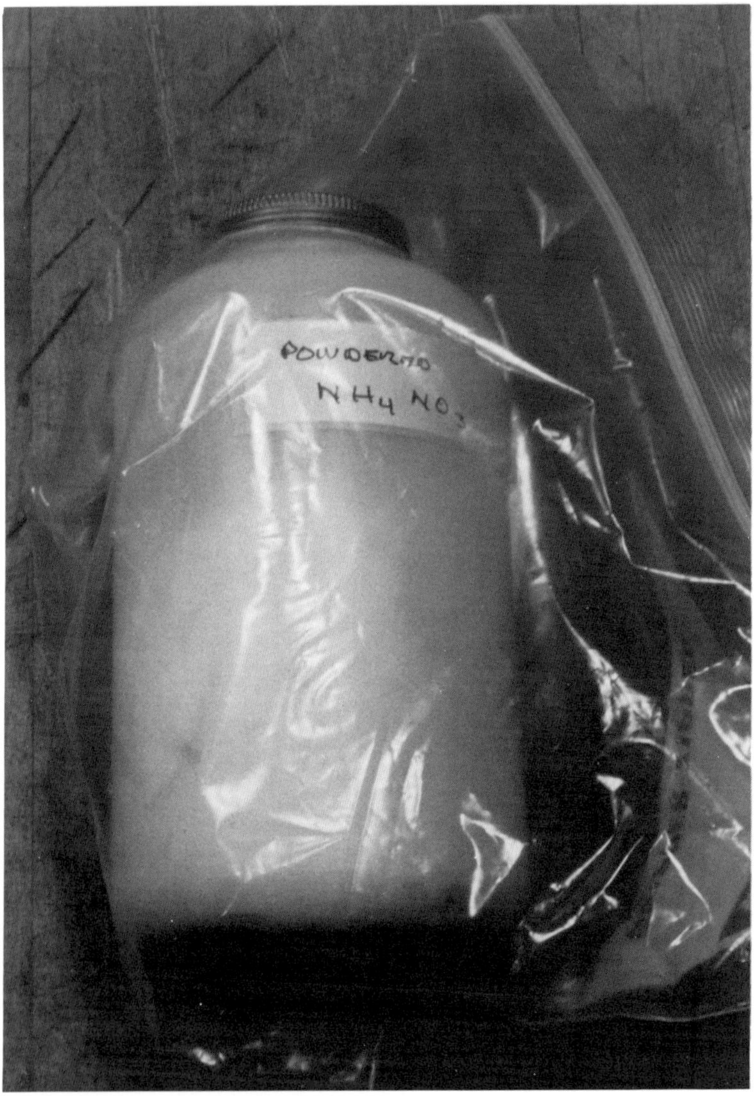

Some improvised explosive makers wash and powder their ammonium nitrate well before they intend to use it. It is then stored in double plastic containers. Yet, in most cases, the ammonium nitrate, which no longer has its protective clay coating, quickly deteriorates because of airborne humidity.

pressure. Some folks report using toilet paper rolls, but plastic plumbing pipe seems to do the job best.

Because of possible internal reactions that might unduly sensitize this material, you cannot enhance this explosive with aluminum powder. One ordnance expert has suggested lining the plastic pipe with thin aluminum foil, but doing so might dangerously sensitize the entire cartridge if it is stored for more than a few days.

Immediately after it has been assembled, this explosive appears to be quite stable. It does not seem to be shock or match sensitive, but this could change very quickly in a matter of days or as a result of acid salts reacting with aluminum foil (if such is used), or simply as a result of sitting around. Explosives do not get old and weak. They get old and unpredictable.

Boost or double cap these charges, being certain to completely bury the cap in the cartridge. Detonation is very showy, with copious smoke but little flash.

Reportedly, these cartridges will propagate, indicating that one could use five or six one-pound charges together to take out a stump or throw out a big rock. Results are subjective but seem to be about on a par with 60-percent dynamite.

As mentioned, this material would be usable in place of commercial dynamite if you could only locate a cheap, certain source of potassium chlorate.

As kids we used a combination of sugar and potassium chlorate as a replacement for gunpowder. It was an effective formula for small-bore cartridges, up to and including .32-20 rifle ammo. Because explosives were easily purchased in local hardware stores, little thought was given to using potassium chlorate to make blasting powder.

At one point we completely ran out of powder. One of my brothers, who had a day job, had a few dollars so he bought two pounds of potassium chlorate. I labored into the night trying to get some powder started for the following days. Any gunpowder we made had to cure a few days, so someone had to get the lengthy process underway if we were ever to have any new powder.

Two pounds of potassium chlorate is a pretty big batch, but I was tired, in a hurry, and not experienced enough to successfully accomplish this sort of work.

Gunpowder formulation requires use of equal parts of household sugar and potassium chlorate. The sugar, in this case, must be warmed evenly till it just melts and then cooled till the potassium chlorate can be safely added (if you can dip your finger in the molten sugar and not get burned it is safe).

I didn't cool the melted sugar enough, and I added in the potassium chlorate too fast without thorough stirring. At first, a few small flashes or sparks showed up as I mixed. Had I stopped adding the potassium chlorate then, continued to stir, and brought down the temperature, everything would have been okay. But, as it was, I dumped the entire two pounds of potassium chlorate into the sugar. It was a gross overload.

My entire pot of sugar powder started to flare and burn, sending huge clouds of noxious, white smoke into the shop. I threw a burlap bag in the stock tank, soaking it, and then threw it over the bowl containing our sugar powder. Enough energy remained in the formulation to summarily eat up the wet bag.

Dad was very upset because I had scorched the workbench and clouded the shop with smoke. He kept looking around to be sure nothing had caught fire. Brother was very upset because his expensive chemicals went up in smoke. They were supposed to go up in smoke, but also perform productive work in the process, which of course hadn't happened. We were out of chemicals, gunpowder, and money, so we were also out of luck—a situation that as kids we got lots of practice contending with.

The lesson I learned that day was to carefully, slowly mix only small batches of powder, especially when making experimental products. Perhaps it was better to learn early in a small way, rather than later when failure could be catastrophic. In that regard, I was probably well served in having gone to work with a contract blaster at age 13. I knew what explosives could and would do. It may have seemed as though I was fooling around with them, but I really wasn't.

3 Hot Tub Explosive

Perhaps it is not entirely foolish to hope that the central government will avoid banning chemicals used by the yuppie crowd to keep their hot tubs functional. It is hoped that they will not be banned till the very last. In this regard, it is probably helpful for you to learn the following formulation—even though the results are a bit wimpy. Even when boosted with ammonium nitrate, results are slightly less than those from 60-percent dynamite, using Ragnar's "how big a hole does it gouge into hard ground" standard.

INGREDIENTS

Calcium Hypochlorite

The principal ingredient—calcium hypochlorite ($CaCl_2O_2$)—is easily purchased in most stores selling supplies to the hot tub crowd. Purchase the highest concentration possible. In no case should this drop below 65- to 70-percent calcium hypochlorite. The cost at $3 per pound is certainly nominal.

Most calcium hypochlorite comes in 450-gram plastic

Chemicals used by the hot tub crowd can be formulated into somewhat acceptable homemade explosives, and the probability is that the chemicals will be among the last taken off the market by an overzealous federal government.

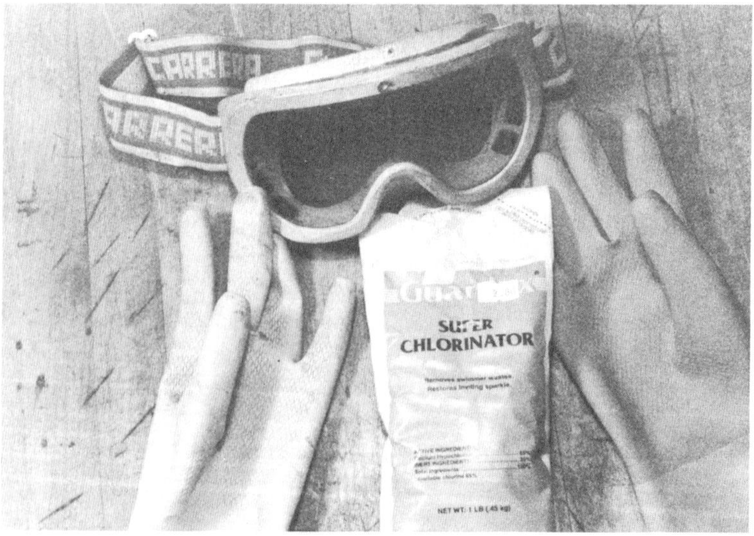

The primary ingredient in the hot tub explosive is calcium hypochlorite, which is used to chlorinate hot tubs. Because calcium hypochlorite is so caustic, great care must be taken when handling this material to protect your hands and eyes.

packets (about one pound). You absolutely must wear a plastic shop-type face shield when handling this material. It is extremely caustic. A single grain of the granular material in the eye could be extremely serious. Rubber gloves are also recommended.

Once the plastic packet is opened, the entire contents must be used. Moisture from even the driest air quickly destroys any chemical left in open packages, even if resealed in airtight plastic containers. In that regard, those who intend to store calcium hypochlorite for any time before using should place unopened plastic packets in airtight glass or poly bottles sealed with plastic (not metallic) lids.

Naphtha

Naphtha, the second ingredient, comes in gallon cans from the local paint store. Unlike somewhat obscure components such as zinc and aluminum dust, virtually all full-service paint stores carry naphtha. It is used to thin some paint and to clean brushes, and sells for about $6 a gallon.

Most conventional formulas out on the street suggest that you can mix about 3-percent naphtha with the calcium hypochlorite to create an explosive. This is no doubt fine advice, with the exception that—as with so many explosives formulas in government manuals—it doesn't work.

MIXING THE EXPLOSIVE

1. Split the 450-gram packet of calcium hypochlorite roughly in half. Place approximately 225 grams in a sound rigid plastic bowl.

2. Quickly add about 60 milliliters of naphtha (predicting the exact amount of naphtha to add under these conditions is difficult). Calcium hypochlorite quickly soaks up the naphtha, forming into clumps of semiviscous sand. Use a wooden,

NEW AND IMPROVED C-4

Put about one-half pound (half a packet) of calcium hypochlorite in a plastic mixing bowl. Note that this is a coarse, granular material.

Using a new wooden paint stick, stir in enough naphtha to form a sandy cohesive mass.

plastic, or glass stirring rod to mix in the naphtha thoroughly. Evaporation seems to occur very rapidly, requiring that you get the naphtha mixed with some dispatch. It is also obvious that the calcium compound has some affinity for the naphtha. It soaks it right up. Use a new wooden paint stick to thoroughly stir till the entire mass has been lightly impregnated with naphtha. Even though naphtha is very inexpensive, adding too much is not advisable. The quality of the explosive seems to decrease rather than increase if the mixture is overly soaked with naphtha.

3. There will be no heat or fumes, other than evaporating solvent, to confound the formulator. The resulting mix seems to be quite stable. No heat, smoke, fire, or whatever give indication of problems. Small amounts of the finished material do not detonate on an anvil as a result of hammer blows.

LOADING THE EXPLOSIVE

To be usable, this clumpy, almost wet material must be quickly packed into solid 1 1/2-inch plastic pipe, cut into six-inch segments; 200 grams will nicely fill one cartridge.

As with many low-grade homemade explosives, this mixture must be detonated by using a booster cap set firmly in the middle of the explosive mass. The effectiveness of this mixture is slightly less than that of 40-percent dynamite.

ADDITIVES

Ammonium Nitrate

Hyping this formula to about 60-percent equivalent is possible simply by adding in about 20 percent by volume of alcohol-washed, dried ammonium nitrate prills. Do not fine-grind the ammonium nitrate. Mix it in as whole prills but keep it very dry.

The calcium hypochlorite-naphtha explosive can be boosted by adding 20-percent (by volume) cleaned, dried ammonium nitrate prills, which have been cleaned in denatured alcohol commonly found in hardware stores.

As with all ammonium nitrate used in explosives, the material must be fresh.

Aluminum Dust

It would seem logical to try increasing the explosiveness by adding aluminum dust, but we concluded that the calcium hypochlorite would rapidly deteriorate the aluminum, thus causing the mixture, which by then really reverts to a compound, to become extremely sensitive. As a result, this is not a recommended procedure even in very small trial quantities. It might work at the moment, but by the next day the compound could become so sensitive that it would blow your hand off just moving it from the shelf. No circumstance now envisioned seems so desperate as to warrant using aluminum dust.

SUMMARY

Quickly summarizing, place 225 grams of calcium hypochlorite in a glass or plastic bowl, and stir in 20 percent, by volume, of prilled, treated ammonium nitrate. Add enough naphtha to just wet the chemical, pack in tight, sturdy cartridges, and fire with a boosted cap. Since this material is very caustic, wear face and hand protection at all times.

A neighbor on the mountain on which I live came storming by one recent evening with what he thought was a virtually insurmountable problem. Seems that while he was out cutting some giant old Douglas fir for firewood, he hung one of the monster trees up on its near neighbor. Cutting the neighboring tree as well would have been dangerous and needlessly wasteful. He didn't need that much firewood. Waiting for winds to eventually drop the tree was not an option because of the danger involved.

As he told it, hikers often came through and they could

end up in a hazardous situation. Besides, he needed stove wood now! As a result I agreed to help to the extent I could. We loaded 20 pounds of powder in his rig and a dozen caps in my truck, both of which we drove up to the leaning tree.

It was really a large one, probably 120 feet tall, 36 inches through the butt, at least 300 years old. Fine old dense wood that experienced people like to use in their stoves. On inspection we found that he had correctly notched the tree. It had broken nicely at the stump, but had fallen only about 10 feet before hanging up. Our first logical goal involved moving the heavy trunk out from the stump, creating more angle that was needed to pull the top down through other adjoining trees. The existing angle of the trunk would not allow us to recut without pinching the saw or risking a dangerous kickback.

The first charge set on top of the stump under the heavy old trunk did lift everything up, moving it back about 10 feet. Now the cut trunk sat on bare earth, but it still didn't fall even though the angle had been increased considerably. We didn't wish to hopelessly fracture the tree trunk, so we chose a less shattering powder in the 40-percent range to complete the job.

Alas, in spite of a showy, noisy blast, we still had a well-hung tree. If anything, it was now more dangerous sitting on soft forest soil rather than on a hard stump. A second six-pound shot moved the log another four feet back and perhaps two feet deeper into the ground. Soft earth blew away from the base of the log rather than lifting and throwing it back. In effect, I was starting to dig a hole into which the tree was dropping.

More powder was going up in smoke than I had originally envisioned, and it was becoming quite dangerous as well as time consuming to work on that tree. The problem resulted from the fact that this tree, although long dead, was so dense and heavy that it actually seemed possible to continue blasting till we had a giant posthole into which the tree would settle.

Our third shot was really a number. We threw in all the remaining powder in a last attempt to move the log to a steep enough angle to fall. Neither of us enjoyed working under the leaning log to dig the powder into place. Packing or stemming the charge was especially difficult because the ground was so light and duffy.

At the shot, duff, debris, rotting tree branches, and dirt flew up in a great cloud. Dead branches shocked off the tree trunk came showering down and would have been deadly had we not elected to stand 50 yards back. There was little noise except what sounded like a 50-pound sledge hitting the ground. In that regard it sounded like a really good shot, one in which explosive energy was contained and directed properly.

Thrown backwards, the tree hung in space for an instant and then fell through the other trees, smartly hitting the ground with a resolute thump. I hadn't planned it so, but hitting the trunk hard broke off many of the dead upper branches, allowing the slick log to fall through without a problem.

In this instance, the tree was down without risk of cutting chunks out of the bottom which could easily have bucked back seriously, hurting the saw operator.

It was a classic use of explosives to handle what would have been a nasty, dangerous situation. As with many explosives applications, the principal limitation was my initial lack of experience.

4 Substitute for Nitromethane

It's the plaintive cry heard throughout the land: "Find something cheaper, easier, and more common than nitromethane with which to formulate our high-velocity, military-grade explosives." Nitromethane, readers say, is tough to find, hideously expensive when finally found, and difficult to store—and, to top it off, it deteriorates in a year or two.

Readers report prices as high as $175 per gallon for nitromethane. A single gallon will mix into roughly 35 pounds of reasonably good high explosive. Nevertheless, this is a very high price to pay unless you are desperate.

As a survival material stored for future use, nitromethane is less than ideal. Recent experience suggests that deterioration starts in earnest after two years, and five or six years later the material is completely dead. The expense of just storing three or four gallons, knowing it is going downhill daily, is a powerful deterrent to many survivors.

There had to be a good substitute for nitromethane out there, but in spite of extensive research on the part of numerous explosives people, any substitute remained elusive.

You could try the century-old expedient of mixing

kerosene or furnace oil with ammonium nitrate. Famous among rock quarry operators, this formula is viable, but the resulting explosive is only equal to about 40-percent dynamite. And this is true only if you pack and contain the explosive well—in holes drilled in solid rock, for instance.

Quarry operators blast with ammonium nitrate but pack it tightly in holes drilled through solid rock.

Lacquer thinner, which in its over-the-counter paint store form contains ethylacetate, acetone, and a number of additional petroleum products, should have been a winner when soaked over ammonium nitrate fertilizer. But it was a complete dud. In spite of using a cap heavily boosted with 50-grain Primacord, nothing happened.

Gasoline or lacquer thinner, into which is thrown several handfuls of Styrofoam peanuts, was suggested by several people, and the idea seemed to have merit. In theory, the

Several explosives experts suggested using gasoline, naphtha, or lacquer thinner to sensitize ammonium nitrate, but results were unimpressive.

thickened solution would have created a rubbery, plastic-like block of explosive similar to genuine, military-issue C-4. Styro peanuts that dissolved in either gasoline or lacquer thinner simply wasted good gasoline or thinner and good packing peanuts. No detonation occurred when this solution was soaked on ammonium nitrate and shot with a boosted cap. Nitromethane, which by itself is a strong solvent, failed to dissolve Styrofoam peanuts.

Had either of these solvents melted paraffin, the solution should, in theory, have both sensitized and stabilized ammonium nitrate. Most improvised munitions books mention liquid paraffin and ammonium nitrate. But purchase of liquid paraffin proved to be elusive, and gasoline and lacquer thinner failed to dissolve it from its block form. This was another excellent idea, at least on paper, that went straight into the tub.

A few years ago, the U.S. media made a great publicity splash over the fact that Czechoslovakian munitions makers were supplying terrorists with a product called Semtex. It seems Semtex was about like regular garden-variety C-4 but with an additional 6 percent vegetable oil blended in. Vegetable oil, the media know-it-alls said, killed the nitrate smell without changing the explosive properties of the explosive very much. Airport bomb dogs couldn't detect Semtex, media wizards reported.

What they didn't report was that Semtex had been around for a long time and that nitrate odors were only somewhat masked by the vegetable oil. Some terrorists admitted that, given a choice, they would have preferred genuine American C-4 for their operations.

Mix ammonium nitrate fertilizer, nitromethane, and 6 percent vegetable oil, one reader suggested: you could have homemade Semtex. As with many of the other suggestions, it was a really fine idea, except not much of a blast resulted.

Obviously, we had to consider other sources of information, no matter how indirect or obscure. After all, such experimentation was how our fraternity got its first C-4 recipe, which has ended up bringing so much joy and satisfaction to readers all over the world. (For those who do not recall the sequence of events that led to the development of C-4, read *Homemade* C-4.)

What, we must ask, do we have commonly available that will substitute for TNT? Ammonium nitrate fertilizer is common as rocks, and we already know where to find aluminum dust.

U.S. government manuals on improvised explosives suggest that common Bullseye powder, familiar to all cartridge reloaders, is cap sensitive and it detonates at velocities of 21,000 feet per second (fps), the same as TNT. Four ounces tested with a cap detonates resolutely, tearing an appropriately sized hole in the ground. In that regard, Bullseye powder, right off the shelf, is a good but extremely expensive substitute for TNT.

According to data published by powder companies ranking smokeless-powder burn rates, Bullseye pistol powder

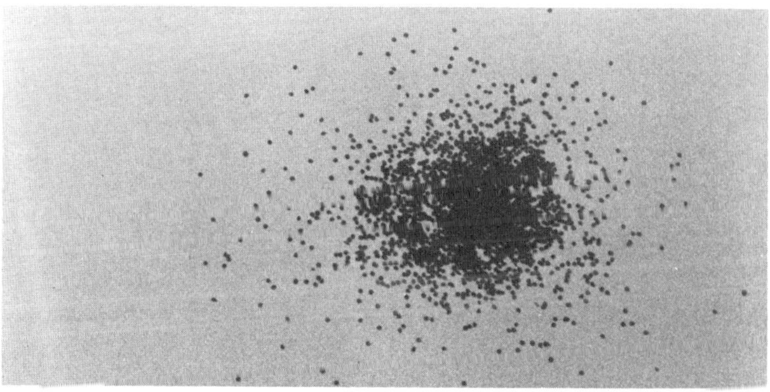

Common Bullseye powder, manufactured by Hercules, can be used in place of nitromethane to sensitize ammonium nitrate.

By using two easy-to-purchase ingredients, Bullseye powder and finely *ground aluminum, makers can produce a product as good as nitromethane-based* C-4, *but with fewer procurement and storage problems.*

manufactured by Hercules is the fastest smokeless powder available to reloaders. Exact ingredients are something of a trade secret, but we now know that Bullseye contains percentages of nitrocellulose, nitroglycerin, ethyl centralize, and rosin, not necessarily in that order.

Physically, Bullseye is formulated into minute, flat, dry flakes. It is sold in one-pound cans and eight-pound kegs. In times past, 15-pound kegs were available, but a combination of shipping regulations and a high price per pound put the kibosh on these. Currently, Bullseye powder sells for about $20 per pound.

When mixed with cleaned, dried, and powdered ammonium nitrate fertilizer at a rate of 16 percent by volume, it detonates the entire mixture very nicely. As with nitromethane-based C-4, any ammonium nitrate used must be fresh, thoroughly dried, washed, dried again, and immediately formulated.

The metal-cutting and cement-breaking ability of this explosive suggests a velocity of about 22,000 fps! When mixed with 3 percent, by volume, aluminum dust, velocities are probably in the 25,000 to 26,000 fps range. As usual, the only gauge is the depth of holes dug in hard earth.

Extensive tests suggest that, at a minimum, you should use 12 percent Bullseye, 85 percent ammonium nitrate, and 3 percent aluminum powder for a dry-powder-mix C-4. All values are computed by volume. More aluminum powder could be substituted for ammonium nitrate, if you have access to a cheap source of this material. But in the overall scheme of things, aluminum powder at about $30 per pound is by far the most expensive component. Lately, we have been purchasing 100 pounds of 34-percent ammonium nitrate fertilizer for $15 ($.15 per pound). Aluminum dust quickly spreads throughout any mixture. As little as 3 percent darkens two to three pounds of ammonium nitrate.

The minimum amount of Bullseye *powder that can be used to sensitize ammonium nitrate is* 12 *percent by volume—except in cases where petroleum jelly is mixed to produce plastique, where a minimum of* 20 *percent by volume must be used.*

A single one-pound can of Bullseye used on the basis of 12 percent by volume would be mixed with about seven pounds of ammonium nitrate and about one-quarter pound of aluminum, yielding a total of about 8.3 pounds of excellent high-grade explosive. The total cost would be close to $3 per pound, or roughly the same as using nitromethane when it sells for $100 per gallon.

However, there are numerous other advantages to using Bullseye powder: shelf life is infinitely longer, storage is simpler, and mixing is easier and can be done well ahead of need. Virtually anyone can find a place to purchase Bullseye powder. Sports shops catering to reloaders are found throughout the nation.

A close substitute for plastique can be made using 20 percent Bullseye powder, 77 percent dried powdered ammonium nitrate, and petroleum jelly.

The mixing of this material should be done in a plastic or glass container. After mixing, it must be stored in a double airtight glass or plastic container. Double resealable plastic bags seem ideal. As with C-4 made by using nitromethane, you absolutely must keep the ammonium nitrate crackling dry or detonation will be wimpy. Even a single drop of water or moisture from a high-humidity day will spoil the ammonium nitrate.

But unlike with some nitromethane C-4 applications, this Bullseye-based material can be wrapped and molded to virtually any shape. Single-strength caps detonate these charges easily and reliably. You do not have to go to the trouble, expense, and danger of working with boosted caps.

Bullseye-sensitized explosive appears to be somewhat impact sensitive, but not dangerously so. No one, to my

As with all ammonium nitrate fertilizer-based explosives, the ammonium nitrate must be oven baked and then washed in new alcohol to remove all moisture. After grinding, the material must be double-sealed immediately.

knowledge, has tested to see if the material deteriorates to a dangerous degree on the shelf, but an educated guess suggests it would not. However, I would still not mix up any explosive that was not going to be used in 30 days or less.

Detonation occurs from high-speed rifle fire, but not from smaller, slower .22 rimfire-type rounds. Small quantities burn about like regular dynamite, which, for those who haven't had to deal with old surplus powder, burns about like paraffin.

A few readers objected to the first C-4 formula, claiming that, although velocity of detonation rivaled that of regular-issue C-4, the physical qualities were not similar.

Those who want something moldable and pliable that closely resembles plastique's physical form can have it with this new improved formula.

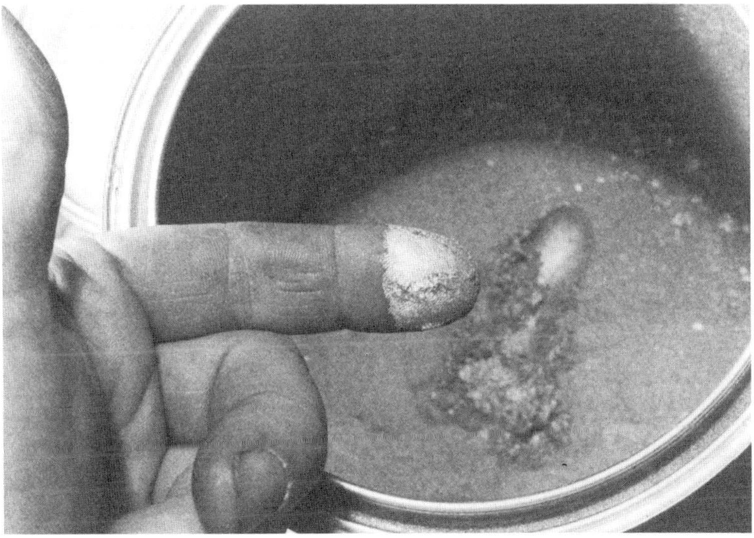

As with old-style nitromethane-based C-4, you can substantially boost the explosiveness of the Bullseye mix by adding in about 3 percent, by volume, of aluminum dust.

Aluminum dust ready to mix into the ammonium nitrate and Bullseye powder.

Start as previously mentioned, combining proper proportions of Bullseye powder, powdered aluminum, and properly treated ammonium nitrate fertilizer. For some unexplained reason, the following formula does not work well with minimum percentages of Bullseye powder, so you must use a minimum of 20 percent Bullseye to achieve acceptable results.

Because aluminum dust permeates the mixture, it quickly darkens the entire batch. You can easily determine whether it is thoroughly mixed by noting its color.

Warm a large jar of petroleum jelly in hot water. Using only enough jelly to bring the mass to desired consistency, spoon in the goo and thoroughly hand knead the mix till a pliable, plastic mass evolves. Generally this is about 15 percent petroleum jelly, by volume.

Petroleum jelly seems not to enter into the reaction, but does create a large plastic mass that can be molded and shaped to exact requirements. Done thoroughly, this mass

Work in enough petroleum jelly so that the entire mixture is plastic and gooey but not sticky.

Working the mixture until it is homogeneous and putty-like. It requires at least 20 percent of Bullseye powder to sensitize the dry ammonium nitrate.

Charges of Bullseye*-based* C-4 *can be molded much like genuine plastique. However, very small charges such as the one shown sometimes do not detonate correctly.*

achieves a balance between sticky, cohesive, and crumbly. The jury is definitely still out on this one, but it appears as though treating with petroleum jelly does at least a partial job of waterproofing the ammonium nitrate. Still, you must store the explosive in double airtight bags, such as the self-sealing plastic types.

The only drawbacks are that (1) it is expensive because of the high percentage of expensive Bullseye required, and (2) charges of under one pound seem not to detonate properly. But in virtually any terms on which you evaluate this explosive, it is superior to old formula C-4. There is also the concern that as people discover this alternative application for Bullseye powder, the powder may disappear from the marketplace.

This small test shot of Bullseye-based C-4 detonated nicely.

Tests continue to determine whether other powder commonly offered to reloaders will substitute for Bullseye. As of this writing, Bullseye seems to be the only game in town.

5 Substitute for Improvisation

Obviously this chapter title is a contradiction. It is appropriate only because of the impossibility of knowing in advance what materials will be easily and commonly available to home workshop explosives makers.

Some readers will, for instance, find acquisition of nitromethane relatively cheap and easy, while others won't be able to find a drop for love or money. Some makers will locate Bullseye powder but cannot purchase ammonium nitrate fertilizer. Materials commonly thought to be obscure will be readily available, or the opposite may be true. A reader recently wrote, saying he could purchase calcium carbide locally for $5 per pound. I can't get fresh calcium carbide, not even for $30 per pound.

It is not my intention to subvert the purpose of this book, which is to outline relatively safe, easy home manufacture of explosives using cheap, easy, relatively common materials. But for those who can lay their hands on the necessary chemicals, in the necessary amounts, the following are excellent explosives. They are highly recommended to do the work, provided you use them within their inherent constraints.

SMOKELESS POWDER EXPLOSIVE

Makers short of Bullseye powder who have only limited access to nitromethane and any other nitrocellulose (smokeless) reloaders' powder should consider the following explosive, a 24,000-fps, military-grade, long-lived plastique based on ammonium nitrate.

(Although most governments throughout the world prohibit private ownership of ammonium nitrate, even for its intended use of making crops grow strong and healthy, published U.S. documents indicate that our authorities are not overly concerned about domestic use of ammonium nitrate in homemade explosives. People do not know how to use it properly in many, if not most, cases. Results are usually

This explosive formulation takes advantage of any nitro-based smokeless powder. Use one level tablespoon of powder (in this case, 5010 powder) mixed into 50 milliliters of nitromethane.

wimpy and are naturally constrained simply because the stuff goes bad so easily and quickly.)

Starting with a common base of ammonium nitrate fertilizer, carefully heat and alcohol-dry 250 milliliters of it. Grind the prills to a fine-powder consistency. Quickly, before the nitrate picks up moisture and sours, mix in 2 percent of aluminum dust. Set this mixture aside in a double-sealed container while preparing the "other half" of the components. As is always true, ammonium nitrate deteriorates with incredibly maddening ease. Only a single drop of moisture, easily rung out of humid environments, will ruin an otherwise fine batch of explosive.

Measure out 50 milliliters of nitromethane and dump in one level tablespoon of any nitromethane-based powder. For purposes of this formula, it need not be Bullseye or any of the faster and perhaps harder to find reloaders' powders. In trials

Dump the measured smokeless powder into the nitromethane and allow it to soften for a few moments.

Stir powder and nitromethane with a plastic rod, creating green-black goo.

Even after an hour of intermittent stirring and soaking, the mixture will still be lumpy.

5010 surplus powder—a very slow, somewhat limited powder originally made for .50-caliber machine gun rounds—was used. Fifty-ten is quite common, easy to work with, and extremely cheap, but any inexpensive, off-brand, smokeless powder will work nicely.

Adding the powder to nitromethane creates a gummy, black gruel. Stir vigorously off and on for about 30 minutes using a plastic or glass rod. The powder will not completely dissolve. It will swell, expand, and soften. Eventually you will have a sticky, dark, green-black cup of gunk.

Thoroughly and completely mix this viscous liquid in with the previously prepared and dried 250 milliliters of ammonium nitrate. A very nice, thick plastic material emerges that can be molded like dark green putty. It can also be loaded in a cartridge or stored in a plastic Ziploc bag. It does not appear

Dump the powder-nitromethane mix into 250 milliliters of thoroughly dried powdered ammonium nitrate.

This mixture is stiff and gooey. Keep turning it over until it is completely blended.

that once it is mixed, the ammonium nitrate mixture will deteriorate very quickly.

There are additional advantages to this explosive besides its using any common smokeless powder and being very easy to detonate. As with genuine plastique, this material is moldable and stable, and its shelf life is reportedly a matter of months. Although these tests were limited to about three weeks, no deterioration or formation of overly sensitive compounds was noted during this time. This formula also uses far less nitromethane than is required for conventional homemade explosives. Those requiring large amounts of explosives might appreciate this feature. Further, moldability suggests that, as with genuine plastique, you can use a minimum of explosive when doing the work.

On the down side, this explosive does require sometimes

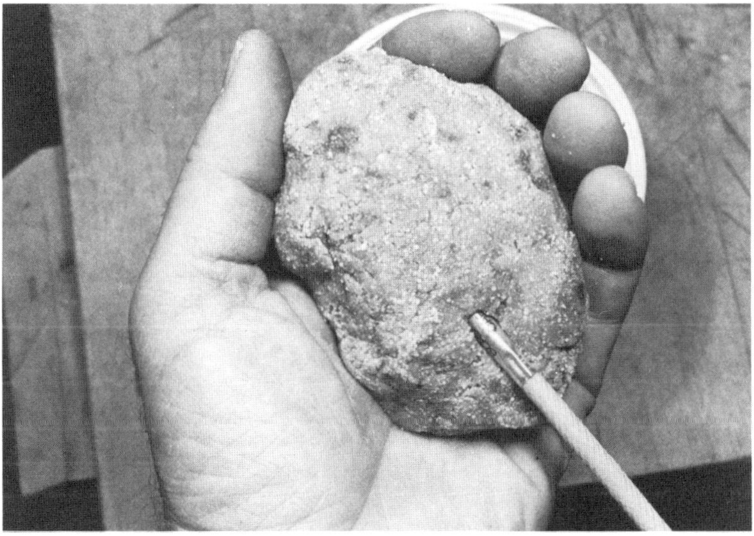

The end result is much like plastique and is relatively stable. It fires with an unboostered cap.

Smokeless powder-sensitized ammonium nitrate is an extremely powerful explosive. Addition of aluminum dust boosts detonation velocity to about 24,000 fps.

tough-to-locate, expensive nitromethane, and one must again contend with fickle ammonium nitrate.

Summarizing briefly:

1. Dissolve a tablespoon of smokeless powder in 50 milliliters of nitromethane. Expect lumpy, black oatmeal-like results.

2. Completely dry 250 milliliters of powdered ammonium nitrate. Add 2 percent of aluminum dust.

3. Combine the two, mixing and kneading thoroughly.

4. Fire with a regular-strength cap.

SUGAR-POTASSIUM CHLORATE PLASTIQUE

A faithful reader raised in a rural community in Texas wrote suggesting that, to him, heaven would be an unlimited supply of potassium chlorate. Compared to when we were kids, potassium chlorate is extremely scarce, and when you do find the chemical it is usuriously expensive. Prices of $20 per pound or more are common. That is probably too much money to constitute a good explosive, no matter how desperate you might be at the moment.

The homemade explosive to follow is probably the easiest to make of any in this book. No washing or drying and very little grinding are required. However, results are about equal to 40-percent dynamite, and costs at $20 per pound for potassium chlorate work out to be about $5 per half-pound stick of explosive.

To build half-pound explosive cartridges one at a time, measure about 75 milliliters of potassium chlorate (75 milliliters works out to be about 167 grams, by weight, or about six ounces).

Tedious as this process is, you must thoroughly pulverize the potassium chlorate crystals. The finished product must

Using a nonsparking stone mortar and pestle, fine-grind 75 milliliters of potassium chlorate.

Measure and mix in an equal volume of common powdered sugar.

have the consistency of face powder, or it will not detonate. Use a stone mortar and pestle to do the work. Pour all 75 milliliters of powder into a clean glass or plastic container.

I do not know the specific gravity of common powdered (confectioner's) sugar, the second ingredient in this simple two-part mix, but it must be less than that of potassium chlorate. However, don't despair; do everything by volume. Use equal parts by volume of both powdered sugar and powdered potassium chlorate. As indicated, this trial test will yield a one-half pound cartridge of high explosive.

Adding the sugar a bit at a time, thoroughly mix into the potassium chlorate. Do this by any dry, clean method possible. Thorough mixing is essential. Some makers roll it in a closed jar. Others pour the two ingredients together and then from container to container.

After mixing, this formula seems very stable and should

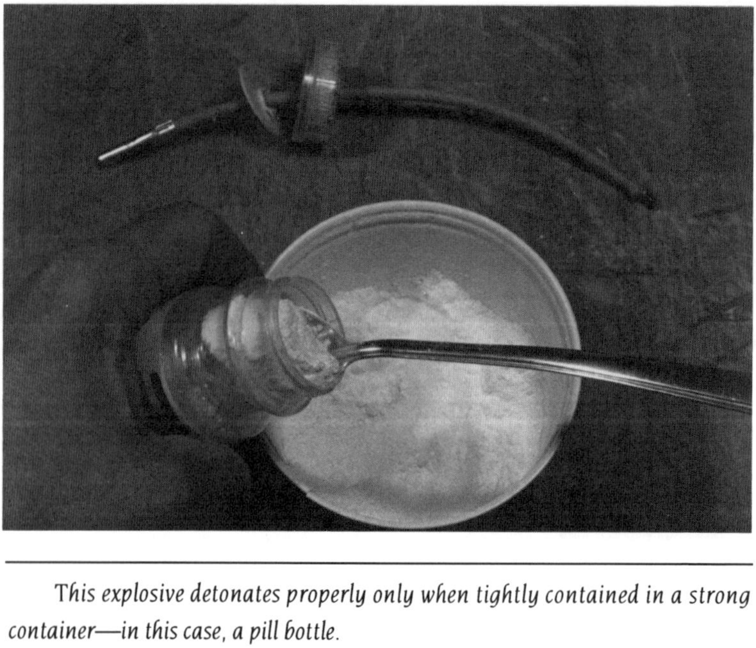

This explosive detonates properly only when tightly contained in a strong container—in this case, a pill bottle.

After spooning all the mixture into the bottle, tamp and press down tightly.

be able to be safely stored away from flame and heat for extended periods. Unlike ammonium nitrate, the material has low affinity for airborne moisture. It is, however, very spark and heat sensitive. Unconfirmed reports claim this mixture burns sufficiently hot to light homemade thermite.

There is one last trick that sets satisfied, knowledgeable explosives users apart from all others. Detonation will not occur unless the explosive is tightly confined in a strong container. Without tight confinement and pressure packing, this explosive will burn fiercely but will not detonate.

Spoon the mixture into a strong plastic container, packing

Single fused caps successfully detonate this mixture.

it in a layer at a time with a wooden dowel or other suitable tool. Fugitive moisture is not a great problem, but you must seal the cartridge after loading. Many makers use large plastic pill bottles onto which they tightly secure the lid.

Undoubtedly, this material could be satisfactorily plasticized using petroleum jelly or perhaps even beeswax. However, it would be an exercise in futility. You would end up with 40-percent plastique, an explosive with limited value. It would not cut steel or crack concrete even if you molded the explosive around an object for a shot. Sugar-potassium chlorate plastique does not detonate unless it is pressure packed and confined.

• • •

I was out of town last spring when a local businessman decided to take out a large, thick footing on a piece of property he was selling. He didn't really know what he was doing and got in a hurry. In retrospect, it seems that I could have saved him some grief.

He vaguely knew about mud-capping rocks to crush them and invalidly supposed that similar physics worked on reinforced concrete. As a result, he foolishly placed six pounds of 60-percent dynamite on the footing, covering them with a great number of old 3x8-inch bridge planks. On top of this he placed two large, expanded steel mats of the type used by the military for emergency runway construction.

His blast was not confined by the makeshift cap of what was really miscellaneous junk. Instead of working on the concrete, the powder worked on the boards and mats.

Besides arousing nearby neighbors, the blast catapulted the mats about 20 feet, and all the planks except one were reduced to matchstick-sized lumber. One remaining plank about 20 feet long went helicoptering off into the air.

Eventually it ran out of lift, coming to rest straight down through the roof of the local Taco Time.

To add insult to injury, the concrete footing was only blackened a bit. No permanent damage resulted—except to the roof of the restaurant. A little knowledge in this case was extremely dangerous. Securing additional information could have produced far better results and would not have sullied the use of high explosives.

As mentioned, you can never know ahead if any of these formulas will be useful at the time you need it. It all depends on your needs and which ingredients are available—something few people can predict with much accuracy.

Conclusion

The original plan for those of you who purchased this book with intentions of discovering how to make several simple explosives using very common materials and to provide dramatic new information regarding a substitute for nitromethane has been accomplished.

In an area where all too many writers seem to be purposefully vague, my goal was to provide detailed, solid, step-by-step information. As is always true, those who find some section of this manual obscure are free to contact me through Paladin for clarification.

Yet please keep in mind that all of this material is for informational purposes only and that working with explosives can be extremely dangerous. I have been working off and on as a powder monkey since the age of 13, which is very close to 50 years. This experience has kept me from making silly, unreasonable errors or taking stupid risks while studying high explosives. Daily observation of the destructive power of high explosives tends to make one extremely cautious.

At this moment, I can easily recall at least two young friends who lost two or more fingers foolishly fiddling around with high explosives about the time I went to work for an old powder monkey. He always stressed caution and safety during the time he schooled me in explosives. He also highlighted the importance of getting some training in chemistry, even if it was only at public high school. He showed me that explosives are only a tool, but one that must be used carefully and accurately.

It hardly seems as though times are sufficiently desperate that people would believe that they must make explosives at home. Perhaps people in rural New York or California who need to construct a road, dig a foundation, or remove some old footings inexpensively might need homemade powder to do the work. But, as a rule, most responsible citizens can still secure sufficient commercial explosives at modest prices with which to perform their desired tasks.

Explosives outlined in this book can be made virtually

Just as safety is uppermost when selecting a fuse, so must safety determine the way in which you use the information contained in this book.

any time. There is no need to get in a hurry and make them now. Keeping "mixed-up" homemade explosives around is more dangerous than keeping commercial explosives on hand—and storing even commercial explosives can be a bit dicey.

In that regard, it would be appropriate if readers actually did treat this volume as a source of future information only. This is the basis on which it was written and sold.

Having fulfilled my contract with readers, it would seem appropriate for purchasers to maintain their end of the bargain.

Ragnar Benson
September 1995